Tony Judt
The Glory of the Rails

ERIS
gems

MORE THAN ANY OTHER TECHNICAL design or social institution, the railway stands for modernity. No competing form of transport, no subsequent technological innovation, no other industry has wrought or facilitated change on the scale that has been brought about by the invention and adoption of the railway. Peter Laslett once referred to "the world we have lost"—the unimaginably different character of things as they once were. Try to think of a world before the railway and the meaning of distance and the impediment it imposed when the time it took to travel from, for example, Paris to Rome—and the means employed to do so—had changed little for two millennia. Think of the limits placed on economic activity and human life chances by the impossibility of moving food, goods, and people in large numbers or at any speed in excess of ten miles per hour; of the enduringly *local* nature of all knowledge, whether cultural, social, or political, and the consequences of such compartmentalization.

Above all, think of how different the world looked to men and women before the coming of the railways. In part this was a function of restricted perception. Until 1830, few people knew what unfamiliar landscapes, distant towns, or foreign lands looked like because they had no opportunity or reason to visit them. But in part, too, the world before the railways appeared so very different from what came afterward and from what we know today because the railways did more than just facilitate travel and thereby change the way the world was seen and depicted. They transformed the very landscape itself.

Railways were born of the industrial revolution—the steam engine itself was already sixty years old when it acquired wheels in 1825, and without the coal that it helped pump to the surface the steam engine could not work. But it was the railways that gave life and impetus to that same industrial revolution: they were the largest consumers of the very goods whose transportation they facilitated. Moreover, most of the technical challenges of industrial modernity—long-distance telegraphic communication, the harnessing of water, gas, and electricity for domestic and industrial use, urban and rural drainage, the construction of very large buildings, the gathering and moving of human beings in large numbers—were first met and overcome by railway companies.

*

Trains—or, rather, the tracks on which they ran—represented the conquest of space. Canals and roads might be considerable technical achievements; but they had almost always been the extension, through physical effort or technical improvement, of an ancient or naturally occurring resource: a river, a valley, a path, or a pass. Even Telford and MacAdam did little more than pave over existing roads. Railway tracks reinvented the landscape. They cut through hills, they burrowed under roads and canals, they were carried across valleys, towns, estuaries. The permanent way might be laid over iron girders, wooden trestles, brick-clad bridges, stone-buttressed earthworks, or impacted moss; importing or removing these materials could utterly transform town or country alike. As trains got heavier, so these foundations grew ever more intrusive: thicker, stronger, deeper.

Railway tracks were purpose-built: nothing else could run on them—and trains could run on nothing else. And because they could only be routed and constructed at certain gradients, on limited curves, and unimpeded by interference from obstacles like forests, boulders, crops, and cows, railways demanded—and were everywhere accorded—powers and authority over men and nature alike: rights of way, of property, of possession, and of destruction that were (and remain) wholly unprecedented in peacetime. Communities that accommodated themselves to the railway typically prospered. Towns and villages that made a show of opposition either lost the struggle; or else, if they succeeded in preventing or postponing a line, a bridge, or a station in their midst, got left behind: expenditure, travelers, goods, and markets all bypassed them and went elsewhere.

The conquest of space led inexorably to the reorganization of time. Even the modest speeds of early trains—between twenty and thirty-five miles per hour—were beyond the wildest imaginings of all but a handful of engineers. Most travelers and observers reasonably assumed not only that the railway had revolutionized spatial relations and the possibilities of communication, but also that—moving at unprecedented velocity and with no impediments to heed their advance—trains were extraordinarily dangerous. As indeed they were. Signaling, communication, and braking systems were always one step behind the steady increase in power and speed of the engines: until well into the later twentieth century trains were better at moving than stopping. This being so, it was vital to keep them at a safe distance from one another and to know at all times where they were. And thus—from technical considerations and for reasons of safety as much as commerce, convenience, or publicity—was born the railway timetable.

It is hard today to convey the significance and implications of the timetable, which first appeared in the early 1840s: for the organization of the railways themselves, of course, but also for the daily lives of everyone else. The pre-modern world was space-bound; its modern successor, time-bound. The transition took place in the middle decades of the nineteenth century and with remarkable speed, accompanied by the ubiquitous station clock: on prominent, specially constructed towers at all major stations, inside every station booking hall, on platforms, and (in the pocket form) in the possession of railway employees. Everything that came after—the establishment of nationally and internationally agreed time zones; factory time clocks; the ubiquity of the wristwatch;

time schedules for buses, ferries, and planes, for radio and television programs; school timetables; and much else—merely followed suit. Railways were proud of the indomitable place of trains in the organization and command of time—see Gabriel Ferrer's painted ceiling (1899) in the dining room of the Gare (now Musée) d'Orsay: an "Allegory on Time" reminding diners that their trains will not wait for dessert.

*

Until the opening of the Liverpool and Manchester Railway in 1830, people did not travel together in large groups. A typical stagecoach held four inside and ten outside. But it was not much used, and certainly not by those with any choice. The wealthy and the adventurous traveled alone or *en famille*—on horseback, in a post chaise or a private carriage—and no one else traveled far or often. But rail travel was mass transit from the very outset—even the earliest trains conveyed hundreds of people—and it was thus important to establish and offer distinctions: by price, comfort, service, and above all the company a voyager was likely to keep. Otherwise the better class of traveler would not come and the poorest would be priced out.

And thus railways established "classes" of travel: typically three, but up to five in the Russian Empire and India. These classes, which gave rise to our modern use of "first class", "second class", etc., for both practical and metaphorical purposes, were reproduced not just in train wagons and their furnishings but in waiting rooms, public bathrooms, ticket offices, restaurants, and all the many facilities provided at stations. In the fullness of time the particular facilities made available to *first-class* travelers—dining cars, club cars, smoking cars, sleeping cars, Pullman cars—reproduced and came to define (in literature, in art, and in design) solid, respectable, prosperous bourgeois life. In their most ostentatious form—typically the long-distance or international train, the Twentieth Century Limited, the Golden Arrow, or the Orient Express—these exclusive facilities defined modern travel as a peculiarly enviable form of cultural ostentation, high style for a privileged minority.

In time, the railways simplified their social stratification into just two classes. In this they reflected the changes after World War I in much of the West, though not always elsewhere. In part this was also a response to competition. From the 1930s, the motorcar was starting to challenge the train as the conveyance of choice for short and even medium-length journeys. Because the car—like its defunct horse-drawn predecessors the post chaise and carriage—was par excellence a *private* vehicle it threatened not just *rail* travel but the very idea of public transportation as a respectable and desirable way to move. As before 1830, so after 1950: those who could afford to do so opted increasingly for privacy. There was no longer either the need or the desire to regulate publicly provided transport with such careful attention to socially calibrated rankings.

*

Trains are about moving people. But their most visible incarnation, their greatest public monument, was static: the railway station. Railway stations—large terminal stations especially—have been studied for their practical uses and significance: as organizers of space, as innovative means of accumulating and dispatching unprecedented numbers of people. And indeed the huge new city stations in London, Paris, Berlin, New York, Moscow, Bombay, and elsewhere wrought a revolution in the social organization of public space. But they were also of unique importance in the history of architecture and urban design, of city planning and public life.

Bringing a railway line into a large town or city was a monumental challenge. Beyond the technical and social issues—the clearance or removal of whole districts (usually the poorest: over two hundred shops, workshops, and churches, together with thousands of tenement homes, were bulldozed to make space for Grand Central Station), the bridging and tunneling past urban and natural obstacles—there was the implication of placing at the heart of an old city a new technology, a substantial edifice, and a steady, daily flow and ebb of many tens of thousands of people. Where should stations be placed? How should they be integrated into the existing urban fabric? What should they *look* like?

The solutions to these questions created modern urban life. From the 1850s (with the building of the Gare de l'Est in Paris) to the 1930s (with the completion of Milan's gargantuan Stazione Centrale) terminal stations from Budapest to St Louis anchored the contemporary city. Their design ranged from Gothic to "Tudorbethan", from Greek Revival to Baroque, from Beaux

Arts to Neoclassical. Some, notably in early-twentieth-century America, were carefully modeled on Rome: the dimensions of Penn Station in New York were calibrated to those of the Baths of Caracalla (AD 217), while the barrel vault ceiling in Washington's Union Station borrowed directly from the transept vaults in the Baths of Diocletian (AD 306).

These massive edifices—which sometimes offered a clue to their nether function but in later years tended to camouflage it, speaking to other urban structures rather than the rail shed behind them—were a source of immense pride for the city and often furnished an occasion to redesign, in fact if not in name, much of the rest of the town. Major European cities—Berlin, Brussels, Paris, London—were reshaped around their railway terminuses, with broad avenues leading up to them, urban subway and tram networks designed to link the incoming rail lines (typically, as in London, in a loose circle with radial spokes), and urban renewal projects keyed to the likely growth of demand for housing generated by the railway.

The railway station became a new and dominant urban space: a large city terminus employed well over one thousand people directly; at its peak Penn Station in New York employed three thousand people, including 355 porters or "redcaps". The hotel built above or adjacent to the station and owned by the railway company employed hundreds more. Within its halls and under the arches supporting its tracks the railway provided copious additional commercial space. From the 1860s through the 1950s, most people entered or exited a city through its railway terminuses, whose size and splendor—whether seen at close quarters or at the distant end of a new avenue built to enhance its significance (the new Boulevard de Strasbourg ending at the Gare de l'Est in Paris, for example)—spoke directly and deliberately to the commercial ambitions and civic self-image of the modern metropolis.

*

As the design of the station made quite explicit, railways were never just functional. They were about travel as pleasure, travel as adventure, travel as the archetypical modern experience. Patrons and clients were not supposed to just buy a ticket and go; they were meant to linger and imagine and dream (which is one reason why "platform tickets" came into being and were very much used). That is why stations were designed, often quite deliberately, on the model of cathedrals, with their spaces and facilities divided into naves, apses, side chapels, and ancillary offices and rituals. As the locus classicus for such winks and nods to neo-ecclesiastical monumentalism, see St Pancras Station (1868) in London. Stations had restaurants, shops, personal services. They were for many decades the preferred site of a city's primary postal and telegraph offices. And above all, they were the ideal space in which to advertise themselves.

The railway poster, the railway advertisement, the brochure—advertising routes, tours, excursions, exotic places and possibilities—came remarkably early in the history of train travel. It was perfectly clear even to the first generation of train managers that they would be creating needs that they alone could meet; and that the more needs they could generate, the greater their business. Within limits the railway companies handled by themselves the business of advertising their wares—most famously in the magnificently stylized Art Deco and Expressionist posters that dominated station walls and newspaper advertisements from circa 1910 to circa 1940. But although they often owned hotels and even steamships, railways were not equipped to manage the full vertical range of services they had opened up, and this business fell into the hands of a new breed, the tour manager or travel agent, of whom the most important by far was the family of Thomas Cook of Derbyshire.

Cook (1808–1892) exemplifies both the commercial energies released by the possibilities of rail travel and the range of experiences to which these led. Beginning with a small family firm organizing Sunday excursion trains for local temperance clubs, Cook accumulated knowledge about trains, buses, and boats, together with contacts in hotels and places of interest: first in Britain, then in continental Europe, and finally across the Americas. Cook and his successors and imitators organized travel itself; indeed, and in collaboration with the railways, Cook and his successors invented the "resorts" to which people might now travel: bookable by Cook and reachable by rail, whether in the mountains, by the sea, or in "beauty spots" freshly identified for the purpose.

But above all, tour organizers furnished information *about* travel. They made it possible for voyagers to imagine and foresee (and pay for) their journey before making it, thereby enhancing the anticipation while minimizing the risk. Cook's brochures, booklets, and advisory guides—advising travelers on where to go, what to expect, what to wear, what to say, and how to say it—were marketed above all in the new railway station outlets opened by newsagents and booksellers. By 1914 Cook had gone to the logical next step of opening branch offices in or next to railway stations and hotels, publishing railway timetables and even underwriting the train cars and facilities provided en route.

*

The illustrations on railway billboards, or on the colorful literature circulated by tour guides and travel agents, capture something else about the railways: their place in modern art, their versatile serviceability as an icon of the contemporary and the new. Artists themselves were never in any doubt about this. From Turner's *Rain, Steam and Speed* (1844) through Monet's *Gare*

Saint-Lazare (1877), Edward Hopper's *Station* (1908), Campbell Cooper's *Grand Central Station* (1909), and on to the classic poster designs of the interwar London Underground (not least Harry Beck's classic map design of 1932, imitated if not emulated in every subsequent railway and subway map the world over), railway trains and stations formed either theme or backdrop to four generations of modern pictorial art.

But it was in the most modern of all the modern arts that the railway was appreciated and exploited to greatest effect. Cinema and railways peaked in tandem—from the 1920s through the 1950s—and they are historically inseparable. One of the first films ever made was about a train—*L'Arrivée d'un train à la Ciotat* (Lumière Brothers, 1895). Trains are a sensual experience: visual and (especially in the age of steam) aural. They were thus a "natural" for cinematographers. Stations are anonymous, and full of shadows and movement and space. Their attraction for filmmakers is not mysterious.

But the sheer range of films that exploit stations, trains, and the prospect or memory of rail travel remains quite striking. No other form of travel has lent itself to international cinema in quite this way: the horse and the motorcar lack the versatility of the train. Westerns and road movies date quickly, and while they had an international market, they were only ever produced in the United States.

It would be otiose to itemize the films that concern or exploit the railways, from *The General* (1927) to *Murder on the Orient Express* (1974). But it is worth reflecting on perhaps the best known of them all, David Lean's *Brief Encounter* (1945), a film in which the station and the train and its destinations do more than just furnish the props and the occasion for emotions and opportunities. The very specificity of the detail (the transcendent authority of the timetable, the configuration of the station and its location in town and community, the physical experience and plot significance of steam and cinders) makes them far more than a setting. The scenes at Carnforth Station, juxtaposed with the domestic life whose tranquility they threaten, represent risk, opportunity, uncertainty, novelty, and change: life itself.

ERIS

86–90 Paul Street
London EC2A 4NE

265 Riverside Dr. #4G
New York, NY 10025

First published in the *New York Review of Books* 2010. Reproduced here with kind permission from the author's Estate.

Printed in Great Britain

ISBN 978-1-912475-83-4

eris.press